# 4th International Conference on CRISPR Technologies 2021



Online
1-3 October 2021

ISBN: 978-1-7138-4838-7

**Printed from e-media with permission by:**

Curran Associates, Inc.
57 Morehouse Lane
Red Hook, NY 12571

**Some format issues inherent in the e-media version may also appear in this print version.**



Printed with permission by Curran Associates, Inc. (2022)

For permission requests, please contact AIChE
at the address below.

AIChE
120 Wall Street, FL 23
New York, NY 10005-4020

Phone: (800) 242-4363
Fax: (203) 775-5177

www.aiche.org

**Additional copies of this publication are available from:**

Curran Associates, Inc.
57 Morehouse Lane
Red Hook, NY 12571 USA
Phone: 845-758-0400
Fax: 845-758-2633
Email: curran@proceedings.com
Web: www.proceedings.com

# 4th International Conference on CRISPR Technologies 2021

Online
1-3 October 2021

AIChE
The Global Home of Chemical Engineers

# TABLE OF CONTENTS

## MAMMALIAN SYNTHETIC BIOLOGY



## POSTER SESSION

# CRISPR-Based Detection of Sars-Cov-2 By Detectr

Authors:
**Broughton, J. - Presenter**, Mammoth Biosciences

At Mammoth Biosciences, we are harnessing the natural diversity of CRISPR for a new purpose: a programmable diagnostics platform, called DETECTRâ„¢, that can identify virtually any genetic sequence of interest. During the COVID-19 pandemic, Mammoth reconfigured its DETECTRâ„¢ platform to identify SARS-CoV-2. Our DETECTRâ„¢ technology uses Mammothâ€™s proprietary Cas enzymes to detect the presence of SARS-CoV-2 after isothermal amplification. These assays leverage the flexibility of CRISPR diagnostics to function in a variety of testing environments including centralized laboratories and in point-of-need environments. This work demonstrated not only the sensitivity and specificity of CRISPR diagnostics, but also the speed at which new molecular diagnostics can be developed using our DETECTRâ„¢ technology. We continue to expand and enhance the functionality of our CRISPR diagnostic platform through the development of new protein systems including members of the Cas12, Cas13, and Cas14 families.

# Engineered CRISPR/Cas12 Systems Enable Point-of-Care Detection of Sars-Cov-2 RNA

Authors:
**Jain, P. - Presenter**, University of Florida
Nguyen, L., Cancer and Genetics Research Complex
Macaluso, N. C., University of Florida
Pizzano, B. L. M., University of Florida
Rananaware, S., University of Florida
Stone, B. T.
Gurijala, J., University of Florida

There is an urgent need for a reliable point-of-care (POC) test that is rapid, sensitive, specific, inexpensive, easy-to-use, and accessible for detecting SARS-CoV-2 RNA. While isothermal amplification-based RT-LAMP tests have been authorized by the FDA under the EUA for POC testing, these tests have poor specificity and cannot be easily developed to discriminate between different variants of concern (VOC) of SARS-CoV-2. Most Class 2 CRISPR/Cas systems, including CRISPR/Cas12 and CRISPR/Cas 13 mediate a nonspecific collateral trans-cleavage of random DNA and RNA after binding or cis-cleavage of their target DNA or RNA. This trans-cleavage activity has been widely exploited for nucleic acid detection, including SARS-CoV-2 RNA. Both Cas13a-based and Cas12a-based assays have been assigned EUA by the FDA for testing SARS-CoV-2 RNA in a high-complexity lab environment, however, FDA has not yet authorized a CRISPR-based POC test for SARS-CoV-2.

We discovered that crRNAs of Cas12a can tolerate various length extensions with DNA or RNA on both 3â€™- and 5â€™-ends (*Nat. Comms.*, 2020). In particular, a crRNA with 7-nt DNA extensions on its 3â€™-end, we named CRISPR-ENHANCE, augmented the rate of Cas12a-mediated trans-cleavage activity by 3.2-fold. By combining with a reverse transcription-loop-mediated isothermal amplification (RT-LAMP) step, a single copy of SARS-CoV-2 RNA could be detected using a lateral flow assay within 50 minutes (*Methods*, 2021). Compared to the wild-type crRNAs, our CRISPR-ENHANCE system demonstrated up to 5-fold higher sensitivity in detecting SARS-CoV-2 RNA. Our CRISPR-ENHANCE-based COVID-19 tests demonstrated high inclusivity and exclusivity and achieved 96.7% sensitivity and 96.7% specificity in clinical samples and a lyophilized version was stable for at least 30 days at room temperature (*medRxiv, 2021; Comms. Med., in rev.)*. We also purified 22 orthologs of Cas12a from different species, and for the first time, combinatorially tested them with 9 crRNA orthologs over 198 combinations and identified several new functional combinations of crRNA/Cas12a (cCRISPR) for SARS-CoV-2 alpha (B.1.1.7) VOC (*medRxiv, 2021; Sci. Adv., in rev.*). Most recently, we combined an ultra-thermo-stable CRISPR/Cas12b with RT-LAMP to create a single-pot reaction that can monitor the RT-LAMP-based amplification and CRISPR-cleavage using a dual-color fluorescence test detecting SARS-CoV-2 VOC within 10-20 minutes in saliva.

# Engineering CRISPR for Rapid Nucleic Acid Detection

Authors:
**Gao, X. - Presenter**, Rice University

We aim to design and engineer a clustered, regularly interspaced, short palindromic repeats-associated system (CRISPR-Cas) to perform ultrasensitive detection of nucleic acids. Our preliminary data has shown that our engineering has significantly improved the CRISPR-Cas properties. By conjugation with other biomedical devices, we aim to develop the next generation platform for nucleic acid biosensing, which is very important for viral detection, including COVID-19.

# Polyvalent Guide RNAs for CRISPR Antivirals

Authors:
**Josephs, E. A. - Presenter**, Duke University
Bagchi, R., University of North Carolina Greensboro
Tinker-Kulberg, R., University of North Carolina Greensboro
Supakar, T., University of North Carolina Greensboro
Chamberlain, S., University of North Carolina Greensboro
Osena-Ligaba, A., University of North Carolina Greensboro
Salehin, M., University of North Carolina Greensboro

CRISPR biotechnologies, where CRISPR effectors recognize and degrade specific nucleic acid targets that are complementary to their guide RNA (gRNA) cofactors, have been primarily used as a tool for precision gene editing but possess an emerging potential for novel antiviral diagnostics, prophylactics, and therapeutics. In gene editing applications, significant efforts are made to limit the natural tolerance of CRISPR effectors for nucleic acids with imperfect complementarity to their gRNAs in order to prevent degradation and mutation at unintended or â€œoff-targetâ€• sites; here we exploit those tolerances to engineer gRNAs that are optimized to promote activity at multiple viral target sites, simultaneously, given that multiplexed targeting is a critical tactic for improving viral detection sensitivity, expanding recognition of clinical strain variants, and suppressing viral mutagenic escape from CRISPR antivirals.

We demonstrate *in vitro* and in living plants (*Nicotiana benthamiana*) that single â€œpolyvalentâ€• gRNAs (pgRNAs) in complex with CRISPR effectors Cas9 or Cas13 can effectively degrade pairs of viral targets with significant sequence divergence (up to 40% nucleotide differences) that are prevalent in viral genomes. We find that CRISPR antivirals using pgRNAs can robustly suppress the propagation of plant RNA viruses, *in vivo*, better than those with â€œmonovalentâ€□ gRNA counterparts. Furthermore, we find that the Cas13â€™s â€œcollateral activity,â€• which is used in CRISPR-based viral diagnostics, can be robustly activated when in complex with an engineered pgRNAs by multiple segments of the RNA genome of SARS-CoV-2 for sensitive detection of the virus responsible for COVID-19.

Despite the advantages for antiviral applications demonstrated here, â€œpolyvalentâ€• guides would be algorithmically rejected using current computational gRNA design tools for gene editing that prioritize specificity in CRISPR targeting. Our computational design pipeline and experimental validation for pgRNAs *in vitro* and *in vivo* demonstrate a powerful new approach to gRNA design for antiviral applications that can be readily incorporated into current viral detection and therapeutic strategies, and they also highlight the need for specific approaches and tools that can address the differential requirements of precision gene editing *vs.* CRISPR antiviral applications in order to mature these promising biotechnologies.

# A More Efficient CRISPR-Cas12a Variant Derived from Lachnospiraceae Bacterium MA2020

Authors:
**Tran, M. - Presenter**, The Scripps Research Institute

CRISPR effector proteins introduce double-stranded breaks into the mammalian genome, facilitating gene editing by non-homologous end-joining or homology-directed repair. Unlike the more commonly studied Cas9, the CRISPR effector protein Cas12a/Cpf1 recognizes a T-rich protospacer adjacent motif (PAM) and can process its own crRNA array, simplifying the use of multiple guide RNAs. We observed that the Cas12a ortholog of *Lachnospiraceae bacterium* MA2020 (Lb2Cas12a) edited mammalian genes with efficiencies comparable to those of AsCas12a and LbCas12a. Compared to these well characterized Cas12a orthologs, Lb2Cas12a is smaller and recognizes a narrow set of PAMs TTTV. We introduced two mutations into Lb2Cas12a, Q571K and C1003Y, that increased its cleavage efficiency for a range of target sequences beyond those of the commonly used Cas12a orthologs AsCas12a and LbCas12a. In addition to the canonical TTTV PAM, this variant, Lb2-KY, also efficiently cleaved target regions with CTTN PAMs. Finally, we demonstrated that Lb2-KY ribonucleoprotein (RNP) complexes edited two hemoglobin target regions useful for correcting common forms of sickle-cell anemia more efficiently than commercial AsCas12a RNP complexes. Thus Lb2-KY has distinctive properties useful for modifying a range of clinically relevant targets in the human genome.

# CRISPR CAS 9

Authors:
**M, V. - Presenter**, AMRITA COLLEGE OF ENGINEERING
CRISPR CAS 9

CRISPR is a technology that can be used to edit genes and, as such, will likely change the world. The essence of CRISPR is simple: it's a way of finding a specific bit of DNA inside a cell.

Cas9 is an enzyme that uses CRISPR sequences as a guide to recognize and cleave specific strands of DNA that are complementary to the CRISPR sequence. Cas9 enzymes together with CRISPR sequences form the basis of a technology known as CRISPR-Cas9 that can be used to edit genes within organisms.

CRISPR gene editing is a genetic engineering technique in molecular biology by which the genomes of living organisms may be modified. It is based on a simplified version of the bacterial CRISPR-Cas9 antiviral defense system. By delivering the Cas9 nuclease complexed with a synthetic guide RNA (gRNA) into a cell, the cell's genome can be cut at a desired location, allowing existing genes to be removed and/or new ones added in vivo.

The technique is considered highly significant in biotechnology and medicine as it allows for the genomes to be edited in vivo with extremely high precision, cheaply and with ease. It can be used in the creation of new medicines, agricultural products, and genetically modified organisms, or as a means of controlling pathogens and pests. It also has possibilities in the treatment of inherited genetic diseases as well as diseases arising from somatic mutations such as cancer. However, its use in human germline genetic modification is highly controversial.

# CRISPR Target Location Determines the Escape Strategy of Bacteriophage

Authors:
Schelling, M. - Presenter, Iowa State University
Sashital, D., Iowa State University
Bacteriophages can evade CRISPR interference by developing mutations in target genomic regions. Here we challenged bacteriophage using Cas12a armed with crRNAs targeting essential and non-essential regions of the lambda phage genome. We find that phage targeted in essential genomic regions mainly develop seed region mutations. However, a mismatch in the seed region is not always sufficient to allow complete phage escape. Mutations that arise depend on the location of existing target mismatches and the sequence context in coding regions. Evidence indicates that the mutations we observe are pre-existing in the phage population at low levels. Phage targeted in non-essential genomic regions develop large deletions that remove the target and allow complete CRISPR escape. Homology at the deletion site indicates that these deletions are generated by homologous recombination, possibly after Cas effector cleavage or during phage genome replication. Our results reveal the importance of natural genetic variation in phage populations for CRISPR immunity.

# Portable Bacterial CRISPR Transcriptional Activation Enables Metabolic Engineering in Multiple Bacterial Species

Authors:
**Kiatissewee, C. - Presenter**, University of Washington
Dong, C., University of Washington
Fontana, J., University of Washington
Sugianto, W., University of Washington
Karanjia, A., Arizona State University
Peralta-Yahya, P., Georgia Institute of Technology
Carothers, J., University of Washington
Zalatan, J., University of Washington
CRISPR-Cas transcriptional programming in bacteria is an emerging tool to regulate gene expression for various applications, including metabolic pathway engineering. Here, we developed and optimized new CRISPR-Cas transcriptional activation (CRISPRa) systems in *E. coli* and *P. putida*. We provide a methodology to transfer CRISPRa to a new host by first optimizing expression levels for the CRISPRa system components, and then applying rules for effective CRISPRa based on a systematic characterization of *E. coli* promoters. We observed a correlation between CRISPR-mediated expression levels in *E. coli* and *P. putida* with $R^2 = 0.80$. The CRISPRa system can be implemented simultaneously with CRISPR interference (CRISPRi) for transcriptional repression, which enables complex regulation of multi-gene pathways. We demonstrated that CRISPRa can be used to regulate heterologous under synthetic promoters and some selected endogenous genes without further modification of the original sequence. We used this optimized system to regulate multi-gene biosynthetic pathways for biopterin and mevalonate production. This work will enable new metabolic engineering strategies in *E. coli* and *P. putida* and pave the way for CRISPR-Cas transcriptional programming in other bacterial species and microbial consortia.

# A Repackaged CRISPR/Cas9 Platform Recasts Non-Homologous End Joining As a Beneficial Instrument in Nonconventional Yeast Engineering

Authors:
Ploessl, D. - Presenter, Iowa State University
Shao, Z., Iowa State University

Inefficient homology-directed repair (HDR) constrains CRISPR/Cas9 genome editing in organisms that preferentially employ non-homologous end joining (NHEJ) to fix DNA double-strand breaks (DSBs). Current strategies to alleviate NHEJ proficiency involve NHEJ disruption. To confer precision editing without NHEJ disruption, we identified shortcomings of conventional CRISPR platforms and developed a novel CRISPR platform, Lowered Indel Nuclease system Enabling Accurate Repair (LINEAR), which drastically enhanced HDR rates (67-100%) compared to previous reports using conventional platforms in four NHEJ-proficient yeasts. With NHEJ preserved, we demonstrate its ability to survey genomic landscapes, identifying loci whose spatiotemporal genomic architectures yield favorable expression dynamics for heterologous pathways. We present a case study that deploys LINEAR precision editing and NHEJ-mediated random integration to rapidly engineer and optimize a microbial factory to produce (S)-norcoclaurine. Taken together, this work demonstrates how to leverage an antagonizing pair of DNA DSB repair pathways to expand the current collection of microbial factories.

# Ginom

Authors:
**Mamun, M. A. A. - Presenter**, University of Alberta
CRISPR CAS 9

CRISPR is a technology that can be used to edit genes and, as such, will likely change the world. The essence of CRISPR is simple: it's a way of finding a specific bit of DNA inside a cell.

Cas9 is an enzyme that uses CRISPR sequences as a guide to recognize and cleave specific strands of DNA that are complementary to the CRISPR sequence. Cas9 enzymes together with CRISPR sequences form the basis of a technology known as CRISPR-Cas9 that can be used to edit genes within organisms.

CRISPR gene editing is a genetic engineering technique in molecular biology by which the genomes of living organisms may be modified. It is based on a simplified version of the bacterial CRISPR-Cas9 antiviral defense system. By delivering the Cas9 nuclease complexed with a synthetic guide RNA (gRNA) into a cell, the cell's genome can be cut at a desired location, allowing existing genes to be removed and/or new ones added in vivo.

The technique is considered highly significant in biotechnology and medicine as it allows for the genomes to be edited in vivo with extremely high precision, cheaply and with ease. It can be used in the creation of new medicines, agricultural products, and genetically modified organisms, or as a means of controlling pathogens and pests. It also has possibilities in the treatment of inherited genetic diseases as well as diseases arising from somatic mutations such as cancer. However, its use in human germline genetic modification is highly controversial.

# CRISPR-Enabled Metabolic Engineering of the Thermotolerant Yeast Kluyveromyces Marxianus

Authors:
**Li, M. - Presenter**, University of California, Riverside
Lang, X., University of California, Riverside
Ramesh, A., University of California, Riverside
Bever-Sneary, D., University of California, Irvine
Alexander, W., Oak Ridge National Laboratory
Trivedi, V., UC Riverside
Baisya, D., University of California, Riveriside
Cheng, J. F., Lawrence Berkeley National Laboratory
Eckert, C., Renewable and Sustainable Energy Institute (RASEI)
Da Silva, N., University of California-Irvine
Lonardi, S., University of California, Riveriside
Wheeldon, I., University of California, Riverside

Engineered CRISPR-Cas9 systems have been developed to enable targeted gene knockouts, gene regulation, and genome-wide screens for functional genomics. The widespread adoption of these systems has rapidly increased access to the genomes of many non-conventional microorganisms, and in doing so has increased the number of hosts that can be engineered for chemical biosynthesis. In this talk, we present our work in developing CRISPR-based genome editing tools for the thermotolerant yeast *Kluyveromyces marxianus* and demonstrate that these tools can rapidly engineer new strains for biochemical production. We focused on *K. marxianus* because various strains have the natural ability to grow at temperatures upward of 50 Â°C and because *K. marxianus* is commonly described as the fastest growing eukaryote, traits that can benefit high temperature, high rate bioprocessing. The ability to metabolize a range of C5, C6, and C12 sugars, as well as organic acids also makes it well-matched with biomass-derived feedstocks. To better exploit these inherent advantages, we have created a set of CRISPR-based metabolic engineering tools for rapid and precise genome editing, and whole-genome screening of novel gene targets for chemical biosynthesis enhancement. Our CRISPR-mediated multigene integration system allowed creating strains that produce high titers of 2-phenylethanol, a valuable flavor and fragrance compound. The genome-wide library of highly active single guide RNAs (sgNRAs) that target every gene in the genome is a promising screening method for functional genomics. It enables rapid genetic mapping of desired phenotypes and guides directed strain revolution. Together, these CRISPR-based genetic tools and examples of their application in metabolic engineering demonstrate that CRISPR systems facilitate strain engineering in non-conventional microbes, and *K. marxianus* is a viable host for chemical biosynthesis.

# Digital Genome Engineering for High-Throughput Discovery and Strain Optimization Applications

Authors:
**Leland, B. - Presenter**, Inscripta
Abbate, E., Inscripta
Krouse, K., National Bioenergy Center, National Renewable Energy Laboratory
Brent, M., Washington University St. Louis
Fox, R., Inscripta
Held, D., Inscripta
Herrgård, M. J., Technical University of Denmark
Kang, Y., Washington University St. Louis
Krishnamurthy, N., Inscripta
Rosains, J., Inscripta
Shepherd, T., Inscripta
Yerramsetty, K., Inscripta
CRISPR-based genome engineering has the potential to accelerate scientific discovery. Current approaches suffer from limitations in scalability, diversity of edit types, and accessibility. Onyx™, a benchtop genome engineering platform, overcomes these limitations by providing the capability to generate up to 10,000 precisely edited and trackable strain variants in *E. coli* in a single run, or up to 6,000 such variants in *S. cerevisiae*. This enables novel approaches to engineer targets within genes, pathways, and even genome wide. The platform simplifies the complex editing workflow, from design to engineered cell library, by including software, reagents, benchtop instrument and analytics. We present data from three applications using Onyx, from high-throughput target discovery to metabolic engineering, to deconvolution of adaptive laboratory evolution variants. In an application to identify new gene targets that improve glycerol utilization in budding yeast, we generated 7 libraries (>40,000 designs) including truncation of non-essential genes and genome-wide expression modulation with terminator or transcription factor binding site modifications. Using barcodes, we identified >100 novel targets with a glycerol utilization phenotype. In a metabolic engineering application, we generated 200,000 designs across 24 genome-edited libraries in *E. coli* and screened the resulting strain variants for improved lysine production. We uncovered more than 15 novel genes outside of the known lysine pathway that contribute to increased lysine production. Rapid recombination of the hits through the Onyx workflow resulted in strains with >10,000-fold increase in lysine. Using Onxy libraries, we deconvoluted variants identified in adaptive laboratory evolution studies, including testing growth in media containing several different inhibitors. The Onyx™ platform will usher in the next era of genomics, enabling researchers to transition from observational studies to precisely engineering living systems at scale. This will have far-reaching benefits for biology, bio-industrial science, agriculture, healthcare, and alternative energy.

# During Meiosis, Gene Expression Is Silenced in Aberrantly Unsynapsed Chromatin and in Heterogametic Sex Chromosomes.

Authors:
**Tzur, Y. - Presenter**, Hebrew University of Jerusalem
During meiosis, gene expression is silenced in aberrantly unsynapsed chromatin and in heterogametic sex chromosomes. Initiation of sex chromosome silencing is disrupted in meiocytes with sex chromosome-autosome translocations. To determine whether this is due to aberrant synapsis or loss of continuity of sex chromosomes, we utilized a novel approach of CRISPR-cas9 engineering in *Caenorhabditis elegans* to create non-translocated, bisected X chromosomes. We analyzed the sequence of the entire X chromosome, identified highly homologous regions within it, and directed four gRNAs to both regions. Using Nanopore long read technology, we screened several lines with small deletions, and identified two strain in which the cuts within the chromosome led to genomic cleavage and aberrant repair that bisected the chromosome. One of the strains contained a linear and circular mini-chromosome was created using DNA vectors, and the other which had only linear segments was created by a RNP approach. In early meiocytes of mutant males and hermaphrodites, X segments are enriched with euchromatin assembly markers and active RNA polymerase ii staining, indicating active transcription. Analysis of RNA-seq data showed that genes from the X chromosome are upregulated in gonads of mutant worms. Contrary to previous models, which predicted that any unsynapsed chromatin is silenced during meiosis, our data indicate that unsynapsed X segments are transcribed. Therefore, our results suggest that sex chromosome chromatin has a unique character that facilitates its meiotic expression when its continuity is lost, regardless of whether or not it is synapsed. Moreover, our work highlight that CRISPR can be used to create new genomic setups that can highlight hidden molecular processes.

# Long-Term DNA Recording with Ordered Insertion Mutations

Authors:
Loveless, T. - Presenter, UC Irvine
Carlson, C., UC Irvine
Ficht, M., UC Irvine
Liang, G., UC Irvine
Liu, C., University of California At Irvine

Mammalian development involves a pattern of disparate but coordinated fate changes undergone by millions to billions of cells over days to years. DNA recording is a technique that has the potential to illuminate these processes at the appropriate scale and depth, by making durable changes in the genome that reflect transient events in the cellâ€™s history. Recently, we published the development of a DNA recorder that writes new DNA, by inserting random nucleotides at a single locus in multiple rounds, such that the order of the nucleotides at the locus reflects the order in which they were added. Our system, called CHYRON (Cell HistorY Recording by Ordered iNsertion), uses Cas9 nuclease, a self-targeting guide (stg)RNA, and the template-independent DNA polymerase TdT. CHYRON records an average of 14.6 bits per recording site over an average of three rounds, but its recording duration is limited because ~25% of edits lead to information-corrupting deletion mutations, which compound over time, and the desired insertion mutations increase the length of the stgRNA, which reduces its activity. Now, **we have created a new DNA recording architecture, peCHYRON, that maintains the strengths of CHYRON â€“ ordered, information-dense recording â€“ but (1) drastically reduces the rate of information loss through off-pathway mutations, (2) is in principle compatible with tens of rounds of recording, and (3) can in principle record at least tens of orthogonal inputs.** In each round of peCHYRON recording, prime editor makes one single-strand nick to introduce an insertion mutation that includes both a â€œsignatureâ€• sequence that constitutes the record and a â€œpropagation sequenceâ€• that enables the next step of recording. In a peCHYRON stable cell line, at least 97% of edits in each round were the desired, information-preserving insertion mutations, and editing efficiency remained constant over at least the first five rounds of recording.

# A Novel CRISPR-Cas9 Mediated Graphene Field Effect Transistor Sensor for the Real-Time Monitoring of Lung Cancer

CRISPR CAS 9

CRISPR is a technology that can be used to edit genes and, as such, will likely change the world. The essence of CRISPR is simple: it's a way of finding a specific bit of DNA inside a cell.

Cas9 is an enzyme that uses CRISPR sequences as a guide to recognize and cleave specific strands of DNA that are complementary to the CRISPR sequence. Cas9 enzymes together with CRISPR sequences form the basis of a technology known as CRISPR-Cas9 that can be used to edit genes within organisms.

CRISPR gene editing is a genetic engineering technique in molecular biology by which the genomes of living organisms may be modified. It is based on a simplified version of the bacterial CRISPR-Cas9 antiviral defense system. By delivering the Cas9 nuclease complexed with a synthetic guide RNA (gRNA) into a cell, the cell's genome can be cut at a desired location, allowing existing genes to be removed and/or new ones added in vivo.

The technique is considered highly significant in biotechnology and medicine as it allows for the genomes to be edited in vivo with extremely high precision, cheaply and with ease. It can be used in the creation of new medicines, agricultural products, and genetically modified organisms, or as a means of controlling pathogens and pests. It also has possibilities in the treatment of inherited genetic diseases as well as diseases arising from somatic mutations such as cancer. However, its use in human germline genetic modification is highly controversial.

# CRISPR-Based Control of Biosynthetic Pathways, Informed By a Quantitative Investigation of Metabolic Burden

Authors:
**Faulkner, I. - Presenter**, University of Washington
Sparkman-Yager, D. W., University of Washington
Kiatissewee, C., University of Washington
Fontana, J., University of Washington
Carothers, J., University of Washington
The heterologous, engineered genetics often used for biosynthetic chemical production share resources with host gene expression. While the hostâ€™s endogenous expression has evolved along with complex regulatory networks to balance supply and demand, or cost and benefit, heterologous expression requires engineered control systems to avoid overly burdening the host. Optimizing the genetic controllers of pathway enzymes, for example, achieves greater likelihood of acceptable balance between cost and production, avoiding reductions in growth rate, population homogeneity, and production titers or circuit performance. Here, we first investigate the metabolic burden of genetic control systems implemented with CRISPR transcriptional activation and T7 RNA polymerase in E. coli, finding that the overhead cost of CRISPR-based control is relatively high, but that expansion of the circuit to several control nodes that share the overhead cost is an inexpensive addition. We next analyze a set of feedback control circuits suggesting that closed circuits can limit excessive heterologous gene expression at a metabolic cost comparable to simpler, and less stable, open circuitry. Finally, we build metabolic pathways producing valuable chemicalsâ€"an aromatic amino acid derivative and a human milk oligosaccharideâ€"making use of CRISPR-based controlâ€™s expandability into broad, open circuits controlling not only the pathway enzymes, but an potentially large array of endogenously-targeted metabolic regulators or genetic biosensors. These applications illustrate the impressive and versatile capabilities of CRISPR-based genetic control when applied to biosynthetic pathways, and are informed by an enhanced understanding of heterologous contributors to metabolic burden in biosynthetic hosts.

# dCas9-Fusion Peptide Library Display and DNA-Templated Peptide Microarray Self-Assembly for Quantitative Protein Binding Assays

Authors:
**Barber, K. - Presenter**, Brigham & Women's Hospital
Shrock, E., Brigham & Women's Hospital
Elledge, S., Brigham & Women's Hospital
CRISPR-based technologies have primarily been deployed for applications in gene editing and DNA/RNA detection. Here, we develop a CRISPR-inspired peptide display platform by fusing customized peptide libraries to nuclease dead Cas9 (dCas9) and pairing each dCas9-peptide fusion with a unique guide RNA (gRNA) barcode. After purification from a single batch of E. coli, each dCas9-peptide fusion within the mixture self-assembles to the position on a DNA microarray surface harboring a sequence complementary to its gRNA barcode. The resulting DNA-templated peptide microarrays are used for antibody binding studies with patient blood samples. dCas9-display may be further adapted for rapid diagnostics or biological materials assembly using DNA scaffolding.

# Multi-Input CRISPR-Based Kill-Switches for Biocontainment of Genetically Engineered Microbes

Authors:
**Moon, T. S. - Presenter**, Washington University in St. Louis
Rottinghaus, A., Washington University In St. Louis
Ferreiro, A., Washington University School of Medicine
Dantas, G., Washington University in Saint Louis

Probiotics have become an effective framework for diagnostic and therapeutic technologies. However, there are safety concerns associated with using genetically engineered organisms for medical applications. Probiotic microbes have the potential to evolve growth advantages over natural microbes and characteristics that are harmful to the host or to the outside environment. To mitigate these concerns, we engineered *Escherichia coli* Nissle 1917 to survive only when and where it is needed using CRISPR-based kill-switches (CRISPRks).

We first designed a CRISPRks that induces cell death by expressing Cas9 and genome-targeting guide RNAs in response to the chemical inducer anhydrotetracycline. This design allows cell killing to occur while the microbe is in the gut in response to oral administration of the chemical. We optimized the efficiency and stability of the CRISPRks and removed the antibiotic dependence for maintenance of the guide RNA plasmid, and deleted genes involved in DNA recombination and mutagenesis. Using this optimized circuit *in vitro*, we achieved more than a 9-log reduction in cell number and demonstrated genetic stability for up to 28 days of continuous growth. This high killing efficiency was maintained *in vivo*, where we achieved complete elimination of the probiotic 24 hours after oral administration of the inducer. This is the first time on-demand elimination of an engineered microbe that has been demonstrated *in vivo*. We next modified our chemically inducible-CRISPRks to also induce cell death in response to ambient temperatures below 33â€™C. This two-input design induces cell killing either in response to oral administration of the chemical or when the microbe is excreted from the body in response to the reduced environmental temperature. This two-input circuit achieved more than a 9-log and 7-log reduction in cell number after exposure to the chemical inducer and temperature downshift, respectively.

Future directions will include incorporating the CRISPRks in microbes engineered to diagnose and treat diverse medical conditions. Our CRISPRks strategy provides a template for future microbial biocontainment circuits. The sensor and killing mechanism employed in the kill-switch are well characterized and functional in many microbes, allowing the CRISPRks design to be broadly utilized.

# Single-Molecule Detection of Sars-Cov-2 RNA Enabled By Plasmonic Sensing of Amplified Nucleic Acids

Authors:
**Ye, H. - Presenter**, University of Texas at Dallas
Qin, Z., University of Texas at Dallas
Bleris, L., The University of Texas at Dallas
Nowak, C., The university of Texas at Dallas
Liu, Y., The University of Texas at Dallas
Li, Y., The university of Texas at Dallas
Single-molecule detection of pathogens such as SARS-CoV-2 is key to combat infectious diseases outbreak and pandemic. Current gold standard for single-nucleic-acid test relies on polymerase chain reaction (PCR). However, its time-consuming protocols and the needs for laboratory infrastructure largely preclude it from point-of-care (POC) testing. Isothermal amplification methods, such as loop-mediated isothermal amplification (LAMP), have emerged as an alternative to PCR and can be performed at POC without the need for thermal cycling. Although simple, LAMP is susceptible to non-template amplification and its by-product-based readouts (e.g., pH change) can produce false-positive results. Previous efforts in advancing the LAMP assay, based on gold nanoparticles (AuNPs)-based sensing of LAMP products, provide simple readouts (e.g., naked-eye detection and later flow strip). However, the weak optical response of AuNPs leads to little sensitivity enhancement, while the non-specific sensing of ionic strength change or molecule-labeled primers limits the detection specificity. Herein, we report an ultrasensitive diagnostic method based on target-specific sensing of LAMP amplicons using gold and silver (Au-Ag) nanoshells as labels, yielding a â€œnaked-eyeâ€• recognition of 10 RNA copies per microliter for SARS-CoV-2 detection. The ultrasensitive detection arises from two distinctive features. First, nanoshells made of Au and silver (Ag) are employed as labels. They have stronger plasmonic extinction in the visible range over AuNPs and provide 20-times sensitivity enhancement in plasmonic sensing. Second, we introduce restrict enzyme digestion and heat denaturation in the detection approach, in order to cut the long concatemers into short repeats that are amendable for subsequent hybridization with plasmonic sensors. In contrast, previous studies relying on direct sensing of LAMP amplicons have met with limited success because of the ampliconsâ€™ stem-loop and cauliflower-like structure. The plasmonic sensing of LAMP amplicons eliminates the contamination from non-template amplification, improving the detection specificity and sensitivity. Our study takes a significant step toward advancing plasmonic sensing to single-molecule detection.

# AUTHOR INDEX

**AIChE**
120 Wall Street, FL 23
New York, NY 10005-4020

ISBN 978-1-7138-4838-7

CURRAN ASSOCIATES INC.
proceedings.com